OUT OF THIS WORLD

Meet NASA Inventor Penny Boston and Her Team's

Martian Cave Colonies

WORLD BOOK

www.worldbook.com

World Book, Inc.
180 North LaSalle Street
Suite 900
Chicago, Illinois 60601
USA

For information about other World Book publications, visit our website at www.worldbook.com or call 1-800-WORLDBK (967-5325).

For information about sales to schools and libraries, call 1-800-975-3250 (United States), or 1-800-837-5365 (Canada).

Produced in collaboration with the National Aeronautics and Space Administration (NASA).

Library of Congress Cataloging-in-Publication Data for this volume has been applied for.

Out of This World
ISBN: 978-0-7166-6261-7 (set, hc.)

Martian Cave Colonies
ISBN: 978-0-7166-6269-3 (hc.)
ISBN: 978-0-7166-6285-3 (pf.)

Also available as:
ISBN: 978-0-7166-6277-8 (e-book)

Staff

Editorial

Director
Tom Evans

Manager, New Content
Jeff De La Rosa

Writer
William D. Adams

Proofreader/Indexer
Nathalie Strassheim

Graphics and Design

Senior Visual
Communications Designer
Melanie Bender

Media Researcher
Rosalia Bledsoe

Acknowledgments

Cover:	© Budkov Denis, Shutterstock
4-13	© Shutterstock
14-15	NASA/GSFC/Arizona State University; Astrobotic Art by Mark Maxwell
16-17	NASA/JPL/UArizona
19-23	© Shutterstock
24-25	NASA
27	© CBS
28-33	© Shutterstock
34-35	© PBS; © Dr. Seth Shostak, Science Source
37	© Ayhan Turan Menekay, Shutterstock
38-39	© Sardo Michael, Shutterstock; Penny Boston
40-41	© Jan Kaliciak, Shutterstock
42-43	© Carsten Peter, Speleoresearch & Films/National Geographic
44	WORLD BOOK photos by Tom Evans

Contents

Glossary There is a glossary of terms on page 45. Terms defined in the glossary are in boldface type that **looks like this** on their first appearance on any spread (two facing pages).

Pronunciations (how to say words) are given in parentheses the first time some difficult words appear in the book. They look like this: pronunciation (pruh NUHN see AY shuhn).

Introduction

Even before the 1969 moon landing, human beings dreamed of visiting the next frontier in space: the planet Mars. Mars calls to explorers with a geologically interesting surface, evidence of water, and most importantly, the possibility of **extraterrestrial life** (life beyond Earth). **Orbiting** satellites, **landers,** and **rovers** are now working around the clock to solve Mars's mysteries. But the planet's deepest secrets may await the arrival of human astronauts.

Martian explorers will have to overcome some incredible challenges. For example, both Earth and Mars orbit the sun at different rates. As a result, good opportunities for travel between the two are fairly limited. Any human visitors would therefore likely have to stay on Mars for at least a matter of months before returning home. The longest that humans stayed on the moon, by contrast, was a little over three days. And, it's likely that humans may one day want to stay on Mars even longer, establishing long-term colonies.

To spend months on the surface of Mars, astronauts will need shelter. And, because cargo capacity is limited, they may not be able to bring much shelter with them. **Astrobiologist** Penny Boston hopes to give future explorers an edge in the Martian housing market. To do so, she's looking to employ a strategy used by some of our earliest ancestors: living in caves.

The NASA Innovative Advanced Concepts program. The titles in the *Out of This World* series feature projects that have won grant money from a group formed by the United States National Aeronautics and Space Administration, or NASA. The NASA Innovative Advanced Concepts program (NIAC) provides funding to teams working to develop bold new advances in space technology. You can visit NIAC's website at www.nasa.gov/niac.

NIAC
NASA Innovative Advanced Concepts

Meet Penny Boston.

❝ Hello, I'm head of NASA's Astrobiology Institute, and I'm interested in caves on distant worlds. I first became interested in such caves as a place to look for extraterrestrial life. Now I'm studying whether they could serve as home to another kind of life—human space explorers. ❞

Mars's desolate surface

The surface of Mars is a barren desert. The planet has a wispy **atmosphere** about one-hundredth as thick as that of Earth. Surface temperatures on Mars rarely climb above the freezing point of water. Harmful **radiation** from the sun and space bombards the planet. Mars is also pelted by tiny bits of rock called **micrometeorites,** grinding surface rock to dust. It is difficult to imagine life developing under such harsh conditions.

Billions of years ago, however, Mars was a different place. It had a much thicker atmosphere that trapped more of the sun's energy, keeping the surface warmer. Much of Mars was probably covered with a shallow ocean. The thick atmosphere and warm ocean may have provided more suitable conditions for living things. It is possible that simple **microbes** arose on Mars.

As surface conditions worsened, such life could have retreated underground.

❚❚ One of the things that we pointed out first in our NIAC work was the potential for the subsurface of Mars to be radically different from the surface—and in many cases, much more likely to be habitable for a much longer period. **❚❚** —Penny

In the subsurface, pressures and temperatures remain just high enough that liquid water may still exist in small quantities.

❚❚ Life could have sheltered in the subsurface as the surface grew cold and dried up. **❚❚** —Penny

Inventor feature:

Studying life that hasn't been found

Boston is an **astrobiologist.** Astrobiology is the study of life in the universe. So far, the only life known is here on Earth. But, astrobiologists are interested in the search for life elsewhere. How do you study something when you're not even sure that it exists?

> ❙❙ My fundamental interests are in the ways that *organisms* [living things] on Earth can make their living in extreme environments. ❙❙ —Penny

Many environments on Earth may seem practically perfect for supporting a variety of living things. But such conditions may be quite rare elsewhere. Luckily, organisms don't just live where conditions seem comfy to us. Here on Earth, life is found in extremes of heat, cold, dryness, darkness, and other harsh conditions. The broader the range of conditions in which life can survive, the more likely we are to find life elsewhere in the universe.

Boston studies **microbes** that survive in extreme conditions: at boiling temperatures, without sunlight, or deep in the ocean. Microbes are such tiny living things as bacteria.

> ❝ Big living things are beautiful and wonderful and magnificent, but they're not fundamental. ❞ —Penny

Microbes are the very essence of life, found wherever larger organisms thrive and many places where they cannot.

How caves form

Orbiting satellites have conducted extensive surveys of the Martian surface. Studying the subsurface can be a lot more difficult. Luckily, Mars provides access to the subsurface—in the form of caves. The surface of Mars is dotted with pits and caves, excellent places to search for subsurface life.

A cave is a natural hollow, or space, in the ground that is large enough for a person to enter. Some caves consist of a single *chamber* (roomlike space) only a few yards or meters deep. Other caves include large networks of passages and chambers.

On Earth, most caves are formed by the action of water as it slowly dissolves underground rock. However, caves can also be formed through volcanic activity. When lava erupts from a volcano, it flows downhill. As the lava transfers its heat to the environment and cools, it begins to *crystallize* (turn solid). But because lava is a complex fluid, different parts of the lava crystallize at different rates.

So you get the formation of tubes of more fluid lava within the overall lava mass. At some point the eruption stops. The edges of the lava flow crystallize, and the more fluid lava in the center drains out, leaving behind these beautiful tubes in the rock. —Penny

Caves around the solar system

When Boston began her research, astronomers didn't know whether caves existed beyond Earth. Many assumed that it took the unique features of Earth's geology to produce caves. But in recent years, **orbiting** satellites using high-definition cameras have cataloged well over a thousand possible cave entrances on the moon and Mars. It is likely that there are ice caves on the frozen moons of Jupiter and Saturn, as well.

The deep shadow of this lunar landform reveals it to be not a crater, but a pit. Scientists have found such pits all over the moon and Mars. Many are likely entrances to cave networks.

Send in the 'bots.

Orbiting satellites can only tell us so much about extraterrestrial caves. The next step may be to send **rovers** to explore these caves. Another NIAC fellow, Red Whittaker, is hard at work developing robotic cave rovers. His project is covered in the volume *Alien Cave Explorers* from the first *Out of This World* series.

This hole is an opening to a lava tube on Mars. Such an opening is called a *skylight*.

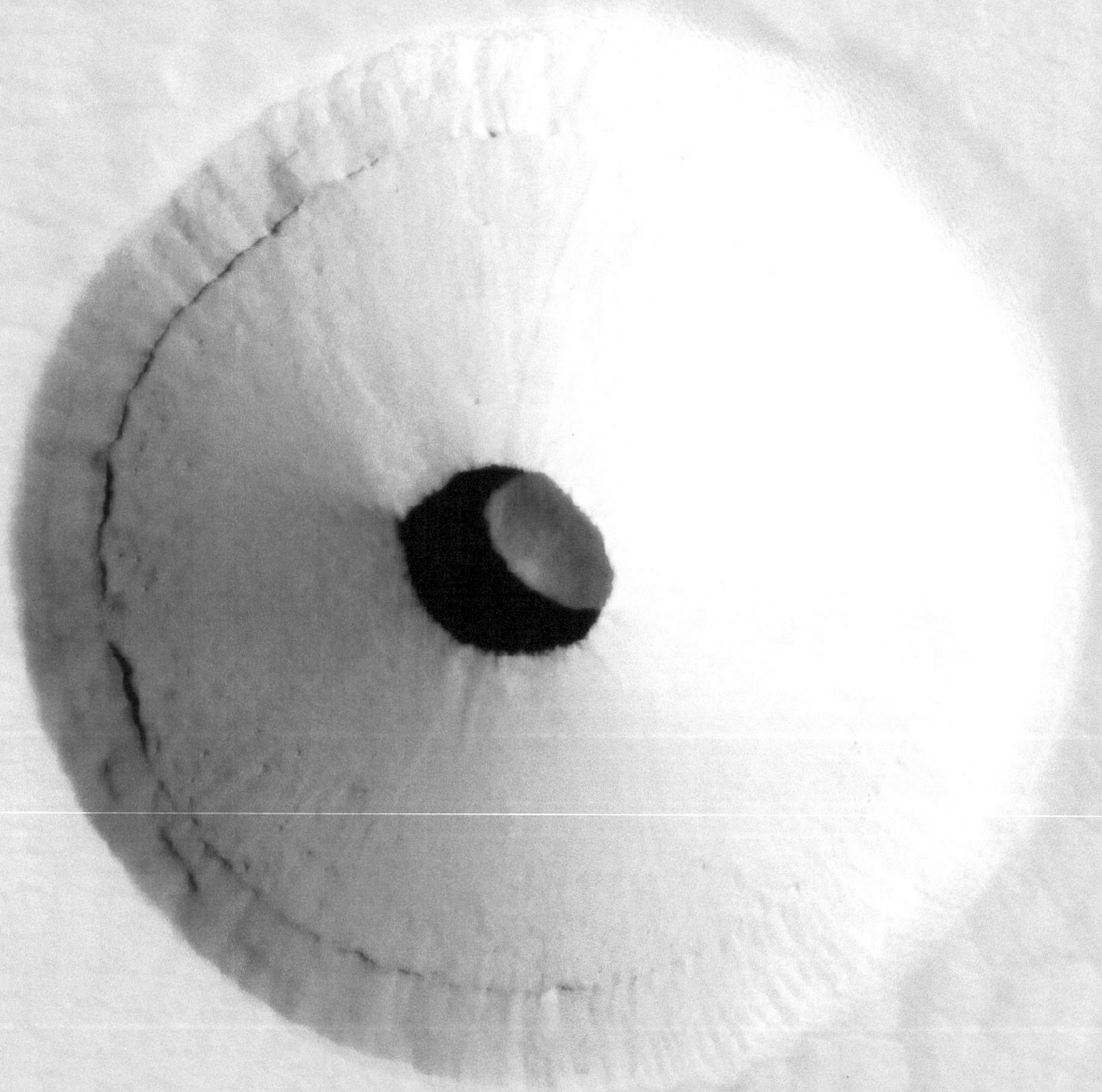

This image (opposite) shows a lava tube on the surface of Mars. The roof of this lava tube has collapsed, but scientists think many intact lava tubes lie beneath the Martian surface.

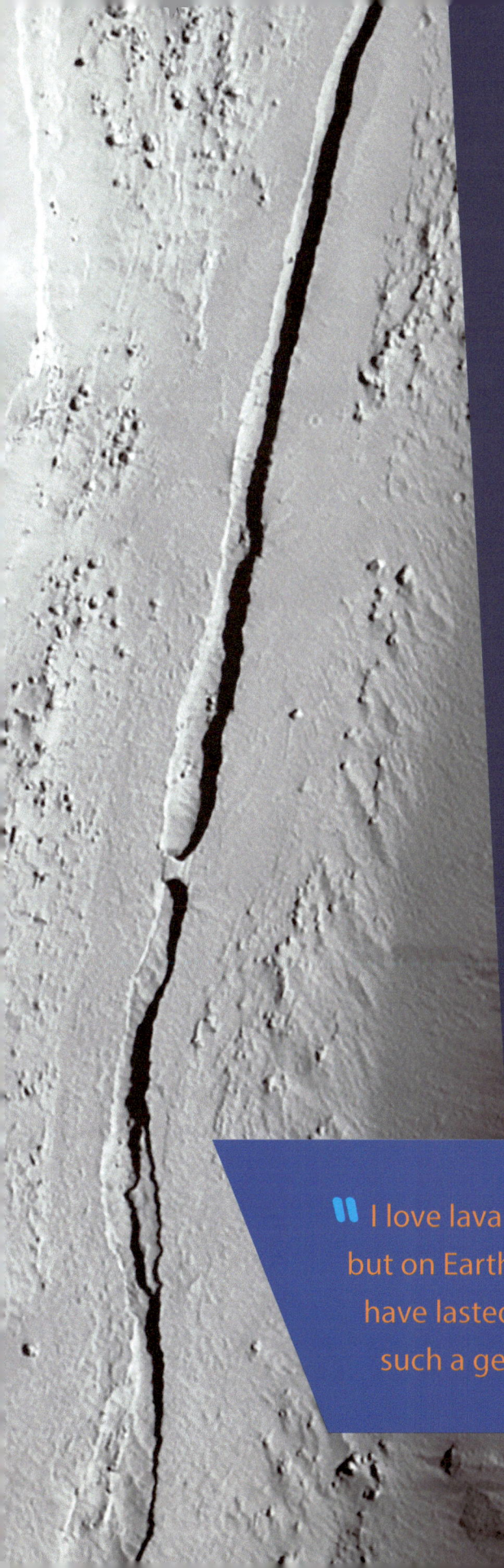

Extraterrestrial caves may have a leg up on their Earthbound counterparts. With plenty of liquid water and volcanic activity, Earth is probably the best place for cave formation in the **solar system.** But, it's the worst at keeping caves around. Earthquakes, water, and other activity also break up and destroy caves. Few caves have formed on the moon or Mars in the last 4 billion years. But, most of them have survived to the present day.

Also, Earth's ample **gravity** bogs down flowing lava and prevents lava tubes from getting too big. On smaller planets or moons, where the pull of gravity is weaker, lava flows can create larger, longer lava tubes. Some lava tubes are big enough to hold entire cities!

❚❚ I love lava tubes, and we have lots of them on Earth, but on Earth they don't last as long as they appear to have lasted on Mars and the moon because we have such a geologically *dynamic* [active] planet. ❚❚ —Penny

Inventor feature:
Falling into caving

As an **astrobiologist,** Boston was interested in how life might arise and *evolve* (change over many generations) on other worlds.

❝ Caves popped up as somewhere we could go here on Earth where we could test some of our ideas—places we could get in relatively cheaply! ❞ —Penny

She and her friend and colleague Chris McKay went on a NASA Ames Research Center expedition to Lechuguilla Cave in New Mexico. They had determined that this cave was similar to those that might be found on Mars.

The expedition was brutal for Boston and McKay, who were both novice cavers. A glob of material dripped from the ceiling of the cave into Boston's eye. Yuck! Her eye became infected,

Lechuguilla Cave in New Mexico

and she had to climb out of the cave with one eye swollen shut. Fortunately, Boston's eye healed after she left the cave and got back into the sunlight.

❝ It nearly killed us. We didn't know what we were doing ... But I fell in love with it after the bruises started to heal! ❞
—Penny

Boston went on to become a skilled caver, exploring caves all over the world and studying the strange life that can be found there.

At home
in the caves of Mars

As Boston studied the possibility of searching for life in caves on Mars, she became convinced that caves would also be a great place for humans to set up shop on the Red Planet. For a temporary base or a long-term colony, caves have several advantages over surface sites.

The sun and other objects in the galaxy produce harmful **radiation** that can damage living cells. Radiation, in fact, is one of the biggest challenges to astronauts and future colonists on the moon or Mars. Radiation exposure from a two- to three-year Mars mission could cause cancer, damage internal organs, and even impact brain function.

Earth's thick atmosphere shields its surface from much of this radiation. Mars has a very thin atmosphere, so harmful radiation bathes its surface. Scientists and **engineers** have yet to develop effective, lightweight radiation shielding to protect astronauts.

But, rock is a great radiation shield. Light is a type of (harmless) radiation—and just think about how little light gets into a cave! An astronaut standing in a lava tube 10 to 15 feet (3 to 5 meters) under the surface on Mars—far enough away from any openings—would be just as safe from harmful radiation as a person standing on Earth's surface.

All that solid rock would also protect astronauts from other threats. Tiny **micrometeorites** constantly bombard the planets and other objects in our **solar system.** Most micrometeorites approaching Earth burn up in the planet's thick **atmosphere.** With only a thin atmosphere, however, the surface of Mars is under constant bombardment.

Micrometeorites may be tiny, but they travel at high speeds, packing a lot of energy. A strike from even the tiniest micrometeorite could cause serious damage to any structures on the Martian surface. Some structures could be fitted with armored panels to limit the damage. But such panels would add weight and take up space, limiting the interior size of a surface shelter. For a shelter constructed in a cave, micrometeorite strikes might barely register as gentle thumps.

A cave is also more *temperate* (moderate) than a planet's surface. Mars is always cold, but temperatures can vary by some 100 Fahrenheit degrees (55 Celsius degrees) over the course of a Martian day. The surrounding rock helps to insulate a cave, stabilizing its temperature. A colony inside a Martian lava tube would still require heating, but its heating needs would be much steadier throughout the Martian day and year.

Caves might also provide easy access to minerals and resources below the surface. Such resources may include water ice.

Icicles form on the roof of
this cave in Kungur, Russia.

Big idea:
Inflatable cave liners

Caves are a great improvement over the Martian surface, but they're not exactly move-in ready. Caves must be open to the outside for Martian explorers to enter them. But then any air pumped in by humans would just escape. Martian colonists might seal the cave opening with an airlock. But many other smaller openings could exist throughout the cave that would have to be closed. Even then, air might seep out through tiny holes in the rock or react chemically with the rock itself.

To prevent such air loss, Boston has proposed lining Martian caves to hold in the air. When astronauts reach a suitable cave, they could inflate a special liner until it fills some or all of the space in the cave. Then, they could coat the liner with special foam that would harden with it, providing more support and protection. Liners and foam could be used to make a huge area habitable but would not take up much **payload** space en route to Mars.

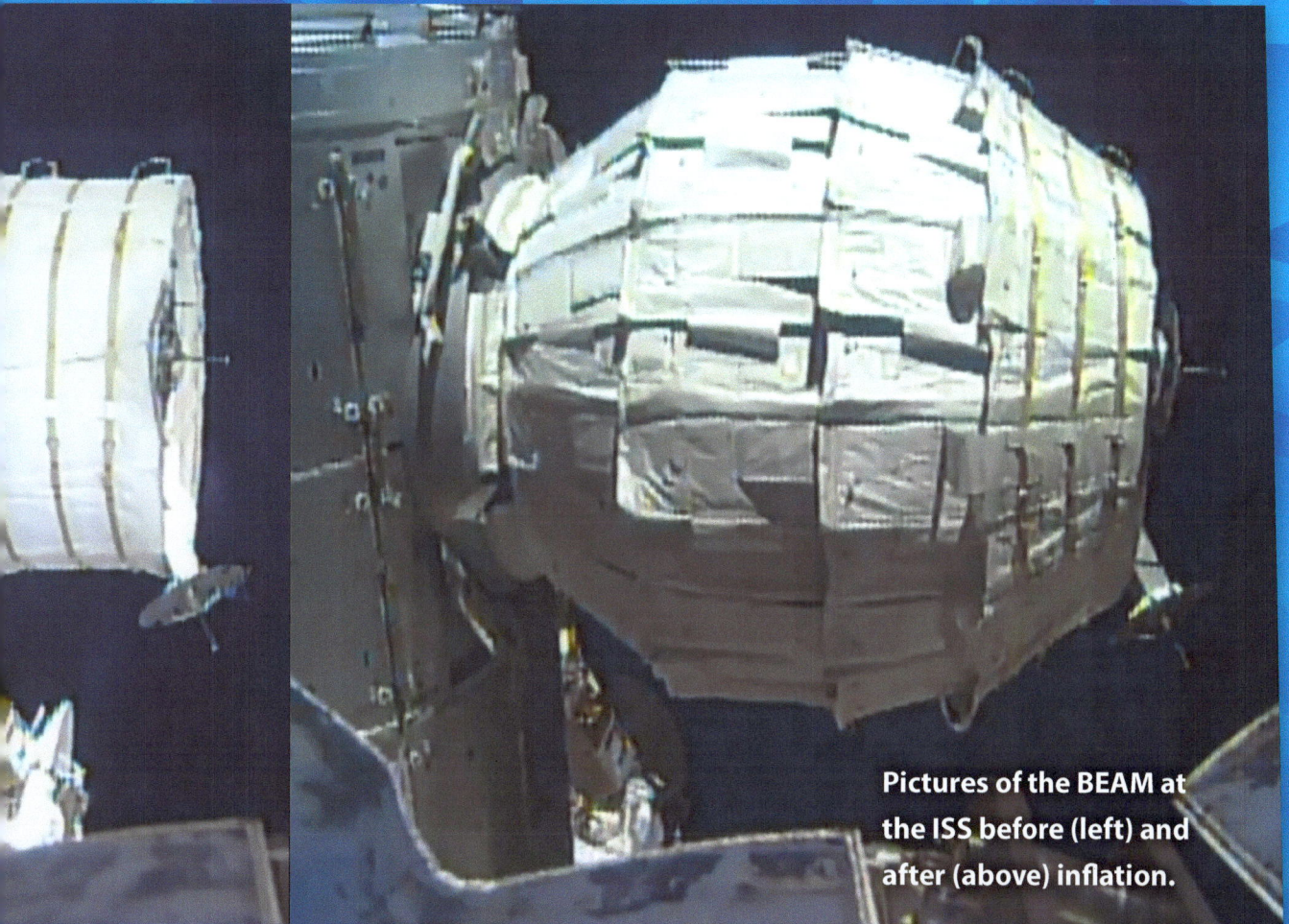

Pictures of the BEAM at the ISS before (left) and after (above) inflation.

Inflatable structures may have a big future
in space. The American company Bigelow Aerospace sent the inflatable Bigelow Expandable Activity Module (BEAM) to the International Space Station (ISS) in 2016. Astronauts attached the BEAM to the ISS, inflated it, and studied conditions inside. With the installation successful, ISS officials decided to keep the module and use it to store extra equipment. Bigelow Aerospace hopes to create entire space stations out of inflatable modules to accommodate manufacturing and tourism.

Inventor feature:
Growing up

Boston grew up during the space race, a period of intense competition in space exploration between the United States and the Soviet Union. The Soviet Union was a country that included all of modern-day Russia along with many surrounding territories. During the space race, the United States prioritized preparing children for careers in science, technology, **engineering,** and mathematics (STEM), grooming them for competition with the Soviet Union.

" Space was in the air! Science was in the air! " —Penny

Boston's parents were not scientists. Her father was an actor, and her mother was a ballerina.

" But, they loved science. They were eternally curious. " —Penny

Boston's parents performed all over the world.

The accompanying travel inspired in Boston a taste for adventure and exploration.

Boston is an enthusiastic Trekkie, a fan of the science-fiction television series "Star Trek" (1966-1969) along with its many spinoffs and sequels. She gave a college commencement speech in Klingon, an invented language of the Star Trek universe. She also enjoyed "Lost in Space" (1965-1968), a show about a family stranded on a starship far from Earth.

❚❚ It was much less sophisticated than "Star Trek." But, the little girl was named Penny, and so am I! ❚❚ —Penny

The cast of "Lost in Space." Penny Robinson, played by Angela Cartwright, is on the far left.

Can we breathe Mars's air?

To inflate cave liners—and, more importantly, to breathe—
Mars colonists will need air. But air is too heavy to bring to
Mars in large quantities. It's true! Air takes up a lot of room
as a gas. It can be compressed into a liquid and stored in
canisters, but such canisters are extremely heavy.

Mars's **atmosphere** is $1/100$ the density of Earth's at sea level.
That's not nearly thick enough for human explorers, who
would need at least $1/3$ the density of Earth's atmosphere at
sea level to remain healthy. But, it would be pretty easy to
compress Mars's atmosphere to higher densities using pumps.

Unfortunately, compressed Martian air would be just as bad for colonists as the uncompressed atmosphere. **Carbon dioxide** gas accounts for 95.3 percent of Mars's atmosphere. The amount of carbon dioxide present in Earth's atmosphere is a fraction of a percent. Humans can breathe a little carbon dioxide, but such a high concentration would be extremely toxic. In addition, oxygen, which makes up 21 percent of Earth's atmosphere, is almost nonexistent in Mars's atmosphere.

Can we breathe *parts of* Mars's air?

Martian air isn't all dangerous, however. The most common gas in Earth's **atmosphere** is nitrogen. Nitrogen gas makes up 78 percent of the atmosphere. It is generally *inert*—that is, it doesn't react easily. We breathe it in and out along with oxygen and other gases. But, aside from a few bacteria, living things don't really make use of this nitrogen gas. If it has a purpose, from our perspective, it is as a sort of "buffer" that keeps more reactive gases from being present in dangerous concentrations.

Nitrogen is the second most common gas in the Martian atmosphere, at 2.7 percent. Argon is the third most common gas, at 1.6 percent. Separating nitrogen gas from argon gas is difficult. But argon is also the third most common gas in Earth's atmosphere, at less than 1 percent. And, it's even more inert than nitrogen gas. Can human beings survive in an atmosphere with less nitrogen and more argon? If it turns out to be possible, we may be able to extract the buffer gases we need from Martian air.

Too much oxygen?

We need oxygen to live, but too much oxygen can be dangerous. *Combustion* (burning) needs oxygen to occur. With more oxygen in the air, things burn more easily and more quickly, with potentially deadly consequences. In 1967, a fire in an Apollo capsule killed three astronauts during ground tests. The capsule had been filled with pure oxygen, which caused the fire to start easily and burn quickly. **Engineers** added nitrogen to the air for future Apollo missions.

Experiments with mouse-tronauts

Boston set out to determine if the combination of argon and nitrogen found in Mars's **atmosphere** could be used as a buffer gas for colonists. She and her team sealed mice in a container with air consisting of 40 percent nitrogen, 40 percent argon, and 20 percent oxygen for several days. A similar mix could be made by compressing the Martian atmosphere, removing the **carbon dioxide,** and adding oxygen. The mice were unaffected, going about their daily routines as normal.

Boston's results are promising. But a larger, longer study will be necessary before the mix can be tested on people. Human explorers would be breathing the mixture for far longer than the mice did in the experiments.

" The mice did very well. And after the experiment, they retired to my house and lived out the rest of their little mousey lives! And they all lived to quite old mouse ages. **"** —Penny

Inventor feature:

Inspirations

When Boston was about 8 years old, she read an issue of *My Weekly Reader* with articles written by Carl Sagan and Frank Drake. *My Weekly Reader* was an educational magazine for children.

Carl Sagan (1934-1996) was an American astronomer, author, and educator. He gained fame as a leading popularizer of science. In Sagan's article in *My Weekly Reader,* he described **astrobiology** and recent advances in the search for extraterrestrial life.

Carl Sagan

Frank Drake (1930-), an American astronomer, is the

leading expert on extraterrestrial intelligence. In 1961, Drake developed a mathematical statement now known as the

Frank Drake

$$N = R_* f_p n_e f_l f_i f_c L$$

Drake equation. The Drake equation provides a framework for understanding the probability that other intelligent civilizations exist in our galaxy. Drake served as a member of the NIAC External Council for many years. In *My Weekly Reader,* Drake discussed his equation and the search for extraterrestrial intelligence.

❚❚ I had already been reading science fiction, so I was already in love with monsters and aliens. ❚❚ —Penny

The magazine hooked Boston on astrobiology. As she continued her education, she was fortunate enough to meet and work with both Sagan and Drake.

Big idea:
Duckweed bioreactor

People and other animals take in oxygen when they inhale. They release **carbon dioxide** when they exhale. In an isolated space, exhaled carbon dioxide would eventually build up to toxic levels. The International Space Station and other spacecraft remove excess carbon dioxide using devices called mechanical scrubbers and filter *media* (materials). But these media need to be replaced periodically. A Martian colony will have to be as self-sustaining as possible, so mission planners hope to avoid dependence on disposable filter media.

Boston has proposed creating a living filter, called a *bioreactor,* that makes use of duckweed to clean the air for the Martian colony. Duckweed is the name of several kinds of tiny plants that float on pools and ponds. A duckweed plant has no ordinary stem or true leaves. The common duckweed is the smallest flowering plant known. It measures only $1/16$ to $3/16$ inch (1.6 to 4.8 millimeters) long.

Frog in duckweed

Like all plants, duckweed makes its own food through **photosynthesis.** It takes in carbon dioxide and water and combines them using light energy to make sugar. Oxygen is released in the process.

Big idea:
Duckweed bioreactor cont.

In a duckweed bioreactor, duckweed would grow in shallow tanks of water. Stale air from the living areas of the colony would be fed into the flat tanks, where the duckweed would absorb some of the carbon dioxide and release oxygen. The clean air would then be returned to living areas. Boston and her team tested such a bioreactor using mouse subjects in a cave.

Some of the duckweed would be harvested from time to time to make

Let there be light.

Plants require light for **photosynthesis,** and people need it to see. Boston has studied creating light tubes to bring surface sunlight down to inhabited caves. Holes could be drilled in the ceiling of the cave, or existing holes could be used. Light would travel from collectors on the surface through highly reflective tubes to *diffusers* (spreaders) in the **habitat** ceiling. The light tubes would be glazed to prevent most harmful **radiation** from entering. LED lights could be used to supplement the hazy Martian sunshine.

room for more to grow. Rather than being thrown away, the harvested duckweed would be put to good use. It could be used as fertilizer for crop plants. Duckweed also has a high protein content, so it could be fed directly to animals or people.

Different air for different fare

A Mars colony will need to be as self-sufficient as possible, so colonists will have to grow much of their own food. It's likely that any colony would be made up of multiple chambers, rather than a single giant room. And, if you have a room that only plants are going to live in, why fill it with an air mixture designed for people?

" The breathing requirements for humans are not the same conditions that plants can tolerate, or even that plants prefer. Plants do quite well in a somewhat higher **carbon dioxide** environment—though this is not good at all for humans. **"** —Penny

Plants need oxygen like people do, so oxygen will still have to be added to make the air breathable. But most plants can handle extra carbon dioxide. In fact, plants could even be genetically modified to use this higher concentration of carbon dioxide to grow more quickly.

" Also, plants can grow at a much lower total pressure. **"** —Penny

Plants can survive at a pressure of just 10 percent that at Earth's surface. The Martian **atmosphere** would thus need to be compressed to just 10 times its regular density. Tailoring the environment in this way would result in valuable savings of precious air. Crop farms could even be tended by specialized robots, keeping humans away from toxic gas mixtures and low pressures.

Protecting
Martian caves

Martian caves could make great homes for future colonists. But remember, these caves may also be the most likely place to find **extraterrestrial life.** So, it will be important for astronauts to proceed with care as they explore and settle Martian caves. **Microbes** from Earth, for example, hitching a ride on future explorers, could contaminate the environment, possibly wiping out any native life before astronauts could even discover it. To avoid this, astronauts will have to sterilize space suits and other gear that will be in contact with the cave environment.

The advantages inherent to Martian caves will make them irresistible to explorers, scientists, and settlers. Boston's work will allow humans to live on Mars decades before technology might be developed to make living on the surface possible.

Cave protection on Earth

Caves are desolate places in which the flow of *organic* (living) matter is greatly restricted. A human wandering through a cave can shed skin, hair, crumbs of food, and other debris. This may not sound like much, but it represents a colossal influx of nutrients into the cave environment.

Such nutrients can totally change the ecology of the cave, favoring aggressive, nutrient-hungry living things over more slow-growing cave species. Studying how human exploration affects cave ecology on Earth can help us to better prepare to protect caves on Mars.

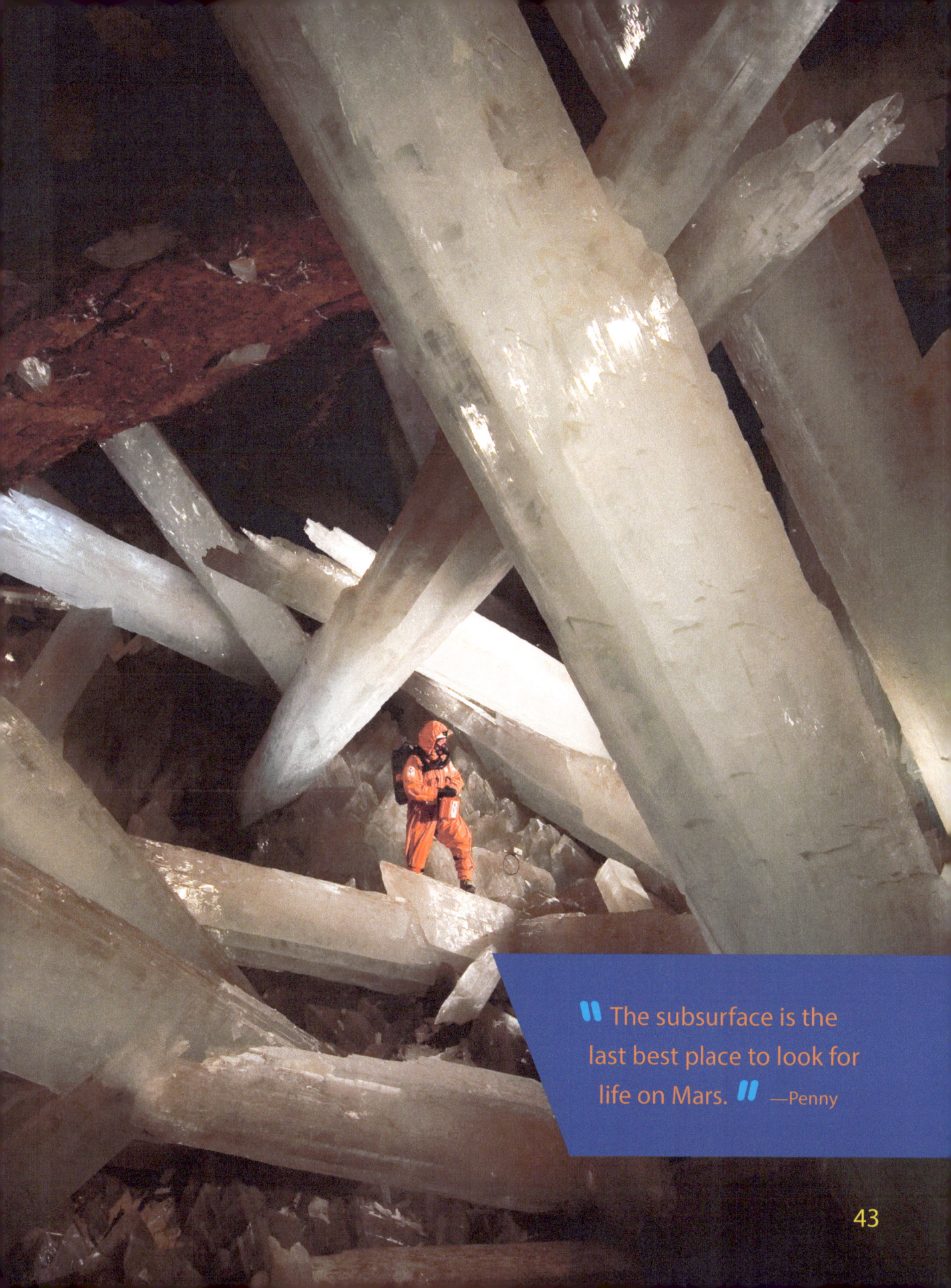

> **❝** The subsurface is the last best place to look for life on Mars. **❞** —Penny

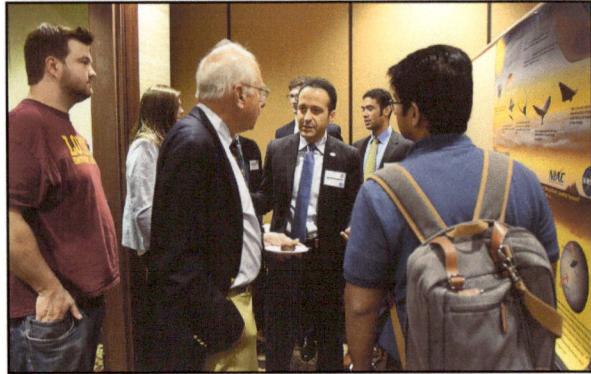

NIAC Symposium

Since NIAC was founded in 1998, it has funded over 350 projects to study next-generation space technology. NIAC holds an annual symposium where the fellows discuss their projects, exchange ideas, collaborate, and visit nearby sites of scientific interest. Penny Boston has been a keynote speaker at the event.

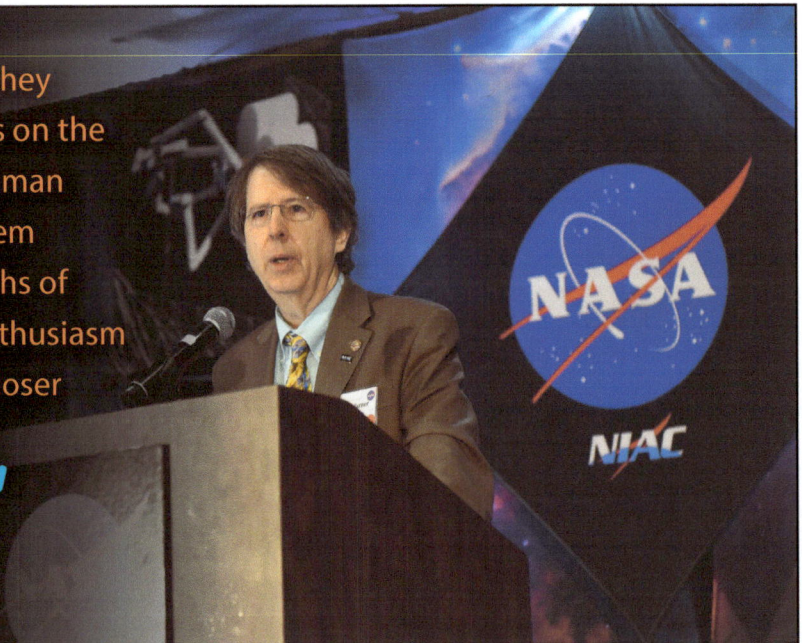

❝ Lava tubes are awesome—they may be the most unique places on the moon or Mars. The future of human colonization of large solar system bodies may well be in the depths of lava tube networks. Penny's enthusiasm and hard work has moved us closer to the day we enter the first of many caves on other worlds. ❞

—Dr. Ron Turner, NIAC Senior Science Advisor, ANSER Distinguished Analyst

Glossary

astrobiologist a scientist who specializes in the search for life on places other than Earth.

atmosphere the mass of gases that surrounds a planet.

carbon dioxide a colorless, odorless gas present in the atmospheres of many planets, including Earth.

engineer a person who uses scientific principles to design structures, such as bridges and skyscrapers, machines, and all sorts of products.

gravitation also called gravitational pull or force of gravity, the force of attraction that acts between all objects because of their mass. Because of gravitation, an object that is near Earth falls toward the surface of the planet. We experience this force on our bodies as our weight.

habitat living area.

lander a spacecraft designed to land on a planet, moon, or other body in space.

microbe a very small living thing.

micrometeorite an object less than 0.04 inch (1 millimeter) wide that strikes a planet or other solar system body.

orbit a looping path around an object in space; the condition of circling a massive object in space under the influence of the object's gravity.

payload the useful load carried by a vehicle.

photosynthesis the process by which plants and other living things make food from carbon dioxide and water.

radiation energy given off in the form of waves or tiny particles of matter.

rover a lander designed to move about for surface exploration.

solar system the sun and everything that travels around it, including Earth and all the other planets and their moons.

Inventor challenge:
Experimental cave

Penny Boston wants people to inhabit caves on Mars one day, but there's plenty more experimental work to be done first. Imagine NASA has just been granted permanent access to a large cave and funds to experiment … and you're in charge! What would you experiment with?

STEP 1

Think about the challenge

Go back through the book and make a list of things that need more research before they can be implemented in a Martian cave colony. Perhaps a couple could be studied together, in the way Boston supplied her mouse-tronauts breathing argon-buffered air with oxygen from a duckweed bioreactor. Think about things not mentioned here, too: how would humans react to living underground for months or years? Write down one or two things you'll be testing for. Those will be your variables.

STEP 2 — Create your prototype

Design your experiment. Be sure to include one or more control groups in your design. In a scientific experiment, a control is a method of isolating the variable you are testing for. For example, a separate group of mouse-tronauts could live in the cave, but breathe a normal air mixture. This would help you determine whether changes in the mice inside the argon environment are due to the gas mixture or something else.

STEP 3 — Share your design

Share your experiment design with friends, classmates, or teachers. See what they think could be improved. If possible, share your experiment with engineers and scientists and ask for their input.

STEP 4 — Grow your idea

There's plenty to research. Design another experiment to go along with your first one. Perhaps you can collaborate with someone else. Maybe you can collect some duckweed and build your own bioreactor!

Index

www.ingramcontent.com/pod-product-compliance
Lightning Source LLC
Chambersburg PA
CBHW052042190326
41519CB00003BA/258